EDI

Purchasing

The Electronic Gateway
to the Future

Steven Marks

PT Publications, Inc.
3109 45th Street, Suite 100
West Palm Beach, FL 33407-1915

Library of Congress Cataloging in Publication Data

Marks, Steven, 1963-
    EDI Purchasing : the electronic gateway to the
future / Steven Marks.
        p. cm.
    Includes bibliographical references and index.
    ISBN 0-945456-27-1
    1. Purchasing--Data processing--Handbooks,
manuals, etc.
    2. Industrial procurement--Data processing--Hand-
books, manuals, etc. 3. Electronic data interchange--Hand-
books, manuals, etc.   I.   Title
    HF5437.M344   1996
    658.7'2'0285--dc20   96-8608

Table of Contents

ABOUT THE AUTHOR: Steven Marks is vice-president—editorial of PT Publications. He has been instrumental in helping authors shape their manuscripts into publishable material. As project editor for PT Publications' series of books on purchasing and other business topics important to the next century, he played a central role in bringing a number of bestselling books to market. Mr. Marks has worked for PT Publications for ten years and was the publisher's original editorial director. Prior to this, he was a free-lance writer and editor, working for major publishers and seminar development companies. Mr. Marks also has extensive experience as a writer and editor with newspapers and national magazines.

Preface

EDI Purchasing: The Electronic Gateway to the Future
provides a brief, but thorough, look at how Electronic Data
Interchange (EDI) is becoming part of the purchasing world.
With this book as a guide, you will be able to build the
foundation for introducing EDI into the purchasing department at your company. And the time to do that is now! The
future is upon us. Electronic networking will introduce
unheard of efficiencies. It will also uncover problems and
that is why we emphasize the connection between EDI and
Business Process Reengineering (BPR) in the purchasing
area.

This book is designed not to be read passively. It is a
book for doers! Pick up a pen or pencil and jot notes in the
margins, answer the questions we pose and write out the
actions you plan on taking.

We take you from an overview of EDI and its uses in
Purchasing to a list of resources which can help you get
your EDI Purchasing project off the ground. Along the way,
you will discover the most important technical issues and
how they can be overcome. Then, you will examine just how
EDI functions in Purchasing and what you will need to

begin your own process of implementation. Since one of Pro-Tech's specialties is in the implementation of systems, we put our focus on this area as well. The best hardware, software and personnel will amount to little or nothing without a solid implementation plan. You will also find out how to select a Value Added Network (VAN) to assist you in connecting to the world of commerce.

We believe this book is the best introduction to EDI Purchasing on the market. And when you combine it with all the other books in our Purchasing Series, we know that you are getting the best education available. Call us for our catalog and list of services. Let us know how you are doing. You just might be in our next book.

Peter L. Grieco, Jr.
West Palm Beach, FL

ELECTRONIC DATA INTERCHANGE

CHAPTER ONE

What is EDI?

EDI (Electronic Data Interchange) is the electronic exchange of business data within a company, between a company and an intermediary network, or directly between two companies. The data is arranged in a structured format so that the sending and receiving computers can "talk" to each other. Most importantly, the structured format allows sending and receiving computers to process the data automatically. For example, Pro-Tech sends a purchase order by EDI to Supplier Y. The computer at Y takes the information and automatically routes it to the appropriate places at that company, such as production scheduling, inventory, finance, accounts receivable, and so on. Once the information is in the structured electronic format of EDI, it does not have to be rekeyed ever again, although companies still retain the ability to change or update the data, some of which would be done automatically as well.

Prior to EDI, the exchange of routine business information such as sales orders, purchase orders and supplier invoices could create 10 to 15 sheets of paper and cost a company between $75 and $125 per transaction. With EDI, companies have been able to lower this cost to as low as 30 cents with, of course, no creation of paperwork. Undoubtedly, there is

Pre-EDI

Typical Purchasing Transaction:

10 to 15 sheets of paper

$75-$125 per transaction

In the EDI Age

Typical Purchasing Transaction:

no paperwork

as low as 30¢ per transaction

somebody who is lowering these costs even further as you read these words. Imagine the savings that could be generated if a company converted 1,000 purchasing transactions per month from paper to electronic transmissions. What about 10,000 transactions? The savings soon add up.

There are reports which indicate that 80% of the information exchanged between companies uses paper as the medium of conveyance. This indicates a tremendous potential for cutting costs. Besides costs, there is also the guarantee of far greater accuracy (no more rekeying), far greater speed of transmission and much improved relations with trading partners which is an inclusive term designating a company's customer, supplier or, as we shall see, an intermediary network.

How does EDI accomplish these savings and improve-ments? Let's follow one transaction to show how EDI works.

1. Operator keys purchase order information into system which creates an electronic purchase order.

2. Electronic PO is sent via EDI to supplier's system.

3. Supplier's system receivesthe electronic PO and automatically sends back an acknowledgment that the electronic form has been received.

4. Supplier's system takes structured data and translates, or reformats, the purchase order information so that it is readable and thus usable.

5. The newly received information electronically enters the supplier's order entry system and is processed. Instructions are sent via internal EDI to productiowarehousing, invoicing, etc. to fill the purchase order.

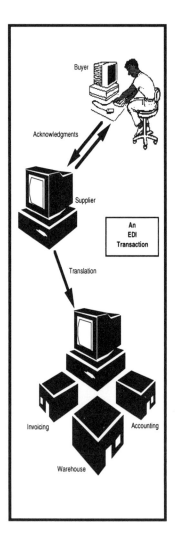

As you can see, EDI builds a connection between the two companies in which routine transactions are handled automatically by the companies' computer systems. This is much the same as instances of EDI in the Finance department where electronic payments and electronic funds transfer are becoming increasingly common.

Frost & Sullivan reports that the market for EDI in the world will grow from $700 million in 1994 to $3.2 billion in 2001, a four-fold increase. To date, EDI has been actively used in three principal areas — commerce, transportation and government. In commerce, banks, financial institutions, airlines and manufacturers using World Class philosophies have led the way. EDI has, for example, made it

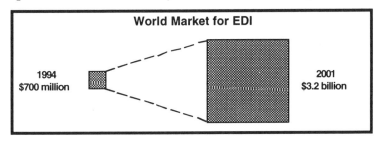

possible now for manufacturers to order and receive supplies with the necessary precision required of JIT. In transportation, EDI has allowed for the almost instantaneous tracking of shipments. Another important development has occurred in the transmission of customs documents via EDI. Activities that once took days now take minutes. And in government, EDI has helped this institution manage and track the vast amounts of information which enter and leave the various departments on a daily basis. U.S. Vice President Al Gore has done much to lead the government in the use of information technologies.

An expanded meaning of EDI which is begining to become almost synonymous with electronic commerce (EC) is that electronic data interchange is the normal business of business conducted electronically, that is, the paperless transaction of business. With this expanded definition, we also begin to touch upon a very significant use of EDI as a tool for reengineering business processes. In a later chapter, we will discuss how EDI can be used to reengineer the purchasing process. EDI is coming to be known as a business enabling technology.

Above all, EDI is fast, inexpensive and accurate.

What is EC and how does EDI fit in?

Electronic Commerce (EC) is a catch-all term for the electronic exchange of data needed to conduct business between trading partners. Whether this exchange is direct or through some intermediary or network, EC has the goal of reducing paper transactions and providing up-to-the-second information. Because the definition is broad, a number of practices can be grouped under the umbrella term.

For example:

- **Bar coding.**
- **EFT (Electronic Funds Transfer).**
- **E-mail.**
- **EDI.**
- **Credit and debit cards.**
- **Computer teleconferencing.**
- **Computer bulletin boards.**
- **Computer forums or SIGs (Special Interest Groups).**
- **Online help desks.**

Such a list cannot be an exhaustive one since the field is growing so rapidly. Much of this growth has, and will be, fueled by the U.S. government's commitment to employ EC in dealings with its enormous supply base. Estimates are that the Department of Defense alone can save over $20 billion in the next few years by implementing EDI with its more than 300,000 suppliers.

Although you probably don't have this many suppliers, even 30 trading partners can present a formidable obstacle to the implementation of EC. Finding the common denominator in terms of software and hardware is an almost impossible problem to solve without common standards. Fortunately, such standards are in place as we shall see in a later chapter. Nevertheless, asking companies to become partners in your EC or EDI program requires patience, hand-holding and practical help. There are companies, for example, which have provided their trading

TRUST

**The Basis of
all Win/Win
Partnerships**

partners with the necessary software or have arranged for them to buy it at a reduced price. The benefits of taking this course will prove more than worthwhile.

Examples of EDI in Purchasing

To give you an idea of how EDI is used in Purchasing, here are some examples from the real world. Kroger, the $2 billion grocery retailer, is a good place to start. The company began to emphasize EDI in 1991 and by 1995 had seen the percentage of grocery purchase orders handled by EDI approach 90 percent. The goal is 100 percent, which will be a remarkable achievement considering the hundreds of suppliers who will be partners in the effort. This expansion is

EDI Pioneer

Kroger

**$2 billion
grocery retailer**

being spearheaded by an EDI team which, among other efforts, is relying upon trading partner symposiums to introduce suppliers to the benefits of EDI. As for the other efforts, Kroger has established a hotline where callers can listen to Kroger's involvement in EDI and ask for implementation guides.

Kroger has also come up with an innovative way of introducing suppliers to EDI. They are asking suppliers to use the computers in Kroger's lobbies to type in their price changes and other product announcements. It is explained that this is just how easy it would be if the computers at their companies were connected by EDI to Kroger. Another first step suppliers can take is an EDI 2 FAX program which allows suppliers who don't have EDI to fax information to a value added network (VAN). The VAN takes the fax and converts it to an EDI message which is then sent to Kroger. Lastly, Kroger will issue EDI kits at negotiated prices to suppliers ready to implement a program.

Another company heavily into EDI is the McJunkin Corporation, a pipe, valve fittings and electrical distributor from Charleston, West Virginia.

EDI Pioneer
McJunkin Corp.
pipe, valve fittings and electrical distributor with 300 suppliers

Their EDI system, called FOCUS, handles purchase orders

to roughly 300 suppliers. McJunkin avoids excessive inventories with a system which creates a purchase order linked to a specific customer order when the item is not in stock. That means that when the item arrives, it immediately gets shipped out without ever being warehoused. FOCUS allows McJunkin to create the purchase order while in the customer ordering side of the system, thus establishing the link.

Inventories can also be dramatically reduced by establishing electronic catalogs on an EDI system. The wholesale pharmaceutical industry has done this with Healthcom® Catalog Services. The service allows manufacturers to upload price changes and product information to the electronic catalog where all of the approximately 240 wholesalers can view the update. In turn, wholesalers can make this information available to their customers — hospitals, drug store chains, independents, etc. The speed of these transmissions is beneficial in itself, but when added to accuracy and cost reductions, this facet of purchasing EDI is of great benefit to a company.

The Benefits of EDI

The benefits of EDI divide into six basic areas. These benefits will vary from industry to industry and organization to organization, but you can expect most of the positive effects. Most importantly, their combined effect will have a significant impact on your company.

1. **Error reduction** — EDI documents are keyed in once or, in many cases, not at all. The information is gathered from company and supplier databases and copied electronically, which means virtually error-free. EDI documents are also stored electronically. No data is corrupted when making copies and then sending them by mail or fax.

2. **Lower labor and processing costs** — It has been estimated that the cost of processing an EDI document represents one-tenth of the cost of the same document on paper. The cost reductions come from lower paper and postage bills, less money allocated to paper office supplies and decreased labor costs since there is less processing of the information. This processing would primarily include items such as keyboarding and data storage and retrieval. The elimination of inefficient manual activities means that less people are needed to do the same amount of work, resulting in the decreased labor costs.

3. **Lead time reduction** — EDI shortens the time between the generation of a purchase order to the fulfillment and delivery of that order. A faster trading cycle is, of course, critical in today's global marketplace which demands agile organizations. Such a company makes or provides a product or service to order, not to stock. We have already explained one system which enables a company to cut the lead time dramatically on delivery of out-of-stock items from its suppliers to the customer. EDI allows stock-outs to be registered immediately and automatically. All of these efforts now make quick response and JIT programs possible on a scale not envisioned before EDI.

4. **Inventory reduction**—This benefit is inseparable from the one above. Shorter lead times mean that less inventory is needed. Signalling a need and checking availability is now a matter of seconds for companies who are connected to their suppliers by EDI. More elaborate systems even allow companies to link directly to a supplier's scheduling department, making it possible to ship products based on a forecast demand which is up-to-date and accurate.

5. **Cash flow** — Now that it is possible to send a PO and obtain instant confirmation of the supplier's ability to fill the order, it is also possible to bill your customers within a much shorter time period. At the same time, these invoices are also virtually error-free, eliminating costly and time-consuming settlements.

6. **Enhanced strategic planning and competitiveness** — Much of the information which is exchanged through EDI can be used as a source of information from which to make marketing and strategic plans. And, the reality is that more and more companies are demanding trading partners who are capable of EDI. Having the capability yourself allows you to retain current customers and make available the same prompt and efficient service to new customers.

The Future of EDI

Any business technology with as many benefits as outlined above is here to stay. EDI's future is also assured because a number of large companies and the U.S. government have made a large commitment to its implementation. And the technological foundation is in place for EDI's

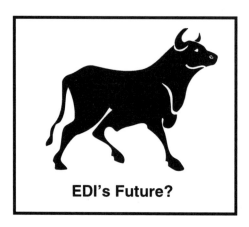

EDI's Future?

continued growth—low-cost computer hardware and software standards. In fact, there are many experts who believe that EDI will pave the way for the virtual corporation. Partnerships will be able to form based on mutual benefit rather than geographic proximity. The speed with which we will be able to transmit information will make this all possible. McJunkin, the distributor mentioned previously, is looking to improve customer relationships in its future EDI development. The company also plans to be heavily involved in the establishment of industry standards which will further help to promote the growth of EDI.

Before you start to paint a rosy picture of the future, you should be aware of some guidelines to keep in mind in order to avoid common EDI problems.

1. **Don't mistake a step toward EDI as the full-blown concept**—Printing out purchase orders on

your EDI system and then faxing or mailing them to your supplier is not EDI.

2. **Don't forget to connect your EDI effort with your business reengineering effort** — EDI is one of the most valuable tools in the reengineering of a business. Don't leave its implementation to a team consisting only of computing or management information specialists. EDI is a strategic tool and needs to be considered from a corporate perspective.

3. **Plan your implementation of EDI** — Avoid allowing circumstances to dictate how and when you implement EDI. A cross-functional team should be looking at where to start and what to do next.

4. **Make quality the number one issue** — The goal of EDI is not only to be fast, but accurate. Implement a system which is well thought out, but also use reengineering to determine if a step or activity is even needed. The less steps, the higher the quality levels.

Keep these guidelines in mind as you prepare for your EDI future. Another key area which must be learned is presented in the next chapter on technical issues.

TECHNICAL ISSUES

CHAPTER TWO

The three technical issues which surface most often in EDI are standards, translation software, and Internet use and security. The first two are well on their way to being fully addressed and much technical progress has been made in the third area. The psychological aspects of security, however, will mean that this issue will take longer to resolve.

Standards

In the early years of EDI, most systems were proprietary. Usually, a large company set standards and instructed all of its suppliers in their use. This worked fine until a need began to grow for communication between different proprietary systems. But, since the systems had different standards, they could not talk to each other directly. Furthermore, the suppliers were beginning to accumulate a new standard for every new customer demanding EDI. A similar situation occurred in bar coding until various industry and government groups set standards. Eventually, this happened with EDI as well. The adoption of format standards now makes it possible for anybody with EDI capability to conduct business with another EDI-capable trading partner.

As early as 1983, standardization began with the ANSI (American National Standards Institute) X12 Committee. Today, the X12 Committee has published standards for over 20 different kinds of business transactions including purchase orders, invoices and requests for quotations. These standards are designed to be generic so that different industries can develop their own subsets. Some of the industries which have done this are automotive, chemical and communications.

Format standards are known as transaction sets which consist of a set of rules for turning business data into specific electronic documents. For example, data elements such as price and quantity are grouped to form a data segment which would be like a line item on a purchase order. A set of data segments then makes up a transaction set. In short, every item on a paper purchase order has an electronic counterpart. Since these transaction sets are the same for each business document, companies now have an agreed-upon format to facilitate their communication.

Now that standards are being accepted, the U.S. government has been able to present one face to all of its suppliers. No matter what department of the government, a supplier will encounter the same transaction sets. Completion of this project is planned for early 1997. Many hope that the government's involvement will create a critical mass which will explode the use of EDI throughout the economy.

What standards are used in your industry or service group?

_____

_____

_____

List the associations you need to contact to gather more
information.

_____

_____

_____

List suppliers who are currently using EDI.

_____

_____

_____

List customers who are currently using EDI.

_____

_____

_____

List the software and hardware used by the above.

_____

_____

_____

Define the quantity of transactions.

_____

_____

_____

Translation Software

EDI translation software begins and ends with the business information in your company's database. For outgoing information, translation software converts the data into the appropriate EDI standard format. The software than establishes a communication link with one or more trading partners and sends a transaction set to their computer systems. For incoming information, translation software converts the information in EDI standard format received from a trading partner into data that can be understood by your company's business applications. For example, a supplier may send updated prices or catalogs to

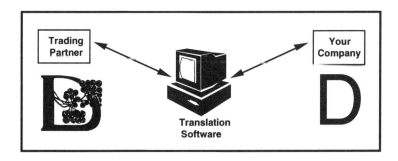

your purchasing department via EDI. This information is received and translated by your software, whereupon the old figures in your database are automatically replaced by the new ones.

All of this translation is accomplished with a system of tables and maps which match data to appropriate transaction sets. In one sense, this system is like an electronic dictionary which translates a foreign language into the native tongue, and vice versa. What you must be aware of is that your translation software fully supports all of the latest industry standards and that it can be easily updated when they change. Off-the-shelf packages do exist for PCs, minis and mainframes. There is an initial purchase cost and an annual maintenance fee for these packages.

Internet and Security

In order for EDI to reach its full potential, it needs to be spontaneous. By that, we mean that you can contact any company or organization in the world and relay electronic business information. This activity needs an inexpensive, reliable and global network. Many believe that the Internet can provide this system to EDI, but all believe that its use will only be possible when security issues are overcome. Various projects are underway to alleviate the concerns of the business world, including the Message Authentification Code (MAC) and public key cryptography. What these and other security methods attempt to do is make sure that information is not lost in the network, that it cannot be altered, that it cannot be read by an unauthorized party, that senders cannot deny sending a message and readers cannot deny reading it, and that an unauthorized party cannot pretend to be an authorized trading partner.

There are many who say that even if these attempts are put into action, the Internet is still too unstructured to be of any real use for business-to-business transactions. Efforts are, however, already being made. CommerceNet, for example, is a consortium funded by government agencies and private companies which has started a number of pilot projects. Other companies are looking into making transactions secure on the World Wide Web. Clearly, the Internet and Web will be part of EDI's future.

List trading partners who currently do business on the Internet or Web.

<u>Company</u>	<u>Contact</u>
_____	_____
_____	_____
_____	_____
_____	_____
_____	_____

BUSINESS ISSUES

CHAPTER THREE

EDI will have both an internal and external impact on your business. The flow of data will change, as will the job descriptions and tasks of a number of employees whose jobs will now entail more computer skills. The manual handling of most documents will be eliminated, but new automated processes will replace it. Many business processes will have to change, and your company will need to adapt to them. These issues will be addressed later in this book. In this chapter, we will take a look at external business issues, your relationships with trading partners.

The amount of benefits you gain from EDI is directly proportional to the number of trading partners you have. That number is, of course, dependent on the relationships that you have built. A win/win partnership is the ideal for which you should strive, but that level of excellence requires a great deal of time and effort. All too often, companies start off with a strong program of mutual benefit only to have it dwindle after a few months. EDI transactions require a high level of trust and confidence, and that can only be obtained by following the guidelines set out below.

Win/Win Partnerships with EDI Trading Partners

In *The World of Negotiations: Never Being a Loser* (0-945456-06-9; PT Publications, West Palm Beach, FL), author Peter L. Grieco, Jr. argues that you must know how to structure a business relationship so that it's a win/win for both sides, so that each party gets what they want. The result will be one success after another. Each of these successes comes without the pressure, anxiety and tension typical of most business relationships.

The approach for win/win requires establishing a trustful and cooperative attitude in order to achieve continuous success. It centers around a philosophy which is based on a total commitment to interaction. Win/win is not a technique. It is a philosophy which consists of several dimensions considered in an evolutionary flow.

DIMENSIONS OF WIN/WIN

Character—Building a foundation for win/win on high integrity, ethics and trust.

Relationship — Building a relationship around a win/win philosophy. Building on the trust which is required to put our cards on the table.

Agreement — Shifting the negotiation to a strategy where both sides can be partners in success through the use of performance or supplier agreements.

Commitment of Management—Getting the company to adopt a win/win philosophy as a way of life.

Process — Creating an environment where negotiation is a process, not a one-time only event. Building a support network for Continuous Improvement Process.

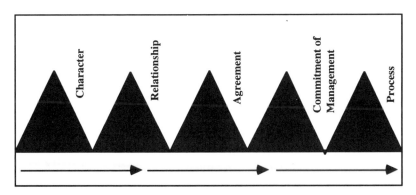

These five dimensions of win/win create the conditions necessary for instilling the habit of interpersonal leadership within the relations between Purchasing and EDI trading partners. Where do you rate your company on a scale of 1 to 100 on each of these axes?

Character

1	25	50	75	100

Relationship

1	25	50	75	100

Agreement

1	25	50	75	100

Commitment of Management

1	25	50	75	100

Process

1	25	50	75	100

Trust and Cooperation

Achieving win/win involves trust and cooperation. As the chart shows, the closer we move to the high end of both axes, the closer we are to achieving win/win. Trust and cooperation are the foundation of win/win and everything else builds upon that foundation. Trust requires integrity and maturity. Stephen R. Covey, author of *The 7 Habits of Highly Effective People,* says that "We have to search within ourselves — beyond scripting, beyond attitudes and behaviors — to achieve validation of win/win."

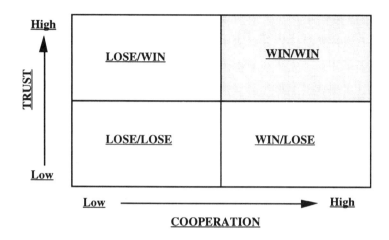

Partnership Agreements with EDI Trading Partners

To solidify the win/win partnership, you should also include paragraphs like the following in your EDI agreement with trading partners:

DECLARATION OF INTENT. Both the Company and the Trading Partner are committed to a philosophy of total customer satisfaction.

This Agreement is entered into with the intention of raising the level of quality and reducing cycle time throughout the whole supply chain, for the mutual benefit of the Parties. The Parties shall work toward achieving zero defects, on-time deliveries, minimal lead times and reduction of costs.

The means for reaching the goals set out above shall be the elimination of activities that do not increase the value for the customer. Other means shall be the development of new processes in cooperation between the Parties and the review of materials, methods of production and administrative routines. Both Parties shall, moreover, internally work towards quality management, reduction of lead time and process improvement.

COOPERATION TEAM. A joint cooperation team shall be formed between the Company and the Trading Partner. It will be led by a Trading Partner's Manager, and comprised by employees from Quality, Manufacturing, Engineering, and Customer Service. Their task is to identify improvements, resolve problems, and implement solutions. The Team would open a line of communication between the Company and the employees of the Trading Partner to allow better understanding of where and how products and processes are used and, in turn, to initiate and drive continuous improvement activities.

The Trading Partner will make customer satisfaction a top priority both for the Company and for the Company's customers.

PERFORMANCE MEASUREMENT. A critical aspect of this Cooperation Agreement are the continuous improvements the Trading Partner implements to satisfy the Company's delivery, quality, cost reduction and service requirements.

The Trading Partner will be rated on these parameters to measure how well the Trading Partner satisfies the Company's requirements. It will be used to track improvements in performance over time and allows the Parties to jointly identify problems and implement corrective actions when necessary. The results from this rating shall be taken into consideration by the Company when assessing future business.

What would you like to add to a cooperation agreement between your company and your EDI trading partners?

_____

_____

_____

_____

EDI IN MANUFACTURING

CHAPTER FOUR

In the manufacturing sector, EDI began as a means of establishing electronic links with first-tier suppliers. First tier suppliers were then often responsible for extending EDI to their partners, and so on down the supply chain. The result was a hodgepodge of proprietary formats. Soon it was discovered that much of the data being sent was basically the same. The automotive industry was one of the first manufacturing industries to start integrating EDI into their operations.

In the early 1970s, automakers were like all other

manufacturers when it came to the use of electronic transactions instead of paper ones. Each automaker had its own private network with proprietary standards. This worked fine for the manufacturers but not for the suppliers who had to maintain a number of different systems. And, since continued business was contingent on accepting the larger company's standards, the smaller suppliers had no choice.

By the mid-1970s, automakers began to see what a problem this was causing and the industry took some action. A pilot program, for example, was started with the trucking industry in which paper bills were replaced by electronic freight bills. By the late 1970s, the Automobile Industry Action Group was formed and had as one its missions to develop standards for data formats and transactions. Although acceptance of the ANSI X12 standard was slow, by the 1980s it had become the accepted standard for transaction sets.

Other manufacturers have followed similar paths to standardization. All can agree that paperwork consumes the resources of time, people and money. In addition, it is slow and inaccurate. This is a situation that the modern lean manufacturer who operates under Just-In-Time (JIT) and cycle time reduction principles cannot tolerate. Most EDI programs at manufacturers today supply at least the following modules:

- **Supplier payments.**
- **Advance shipping notices.**
- **Invoices.**
- **Production forecasts.**
- **Shipping schedules.**

And where once EDI was a program between manufacturers and suppliers, it now includes transactions between manufacturers and their customers. In a short time, EDI has come a long way. It is now being thought of as infrastructure that no company can be without.

EDI and JIT

Before JIT, manufacturers stockpiled parts and subassemblies in order to keep production lines supplied. The problem with this was that inventories represented a big drain on company resources, and the cost of maintaining inventory was substantial. JIT was adopted to bring parts and materials from suppliers into the plant (and, ideally, to the production line) just as they were needed. It soon became apparent that the required coordination of JIT deliveries could not be accomplished without JIT information.

Today, even customers are part of this loop. A purchase at a retail site can now trigger action at a manufacturer via the transmission of purchase data to production scheduling. The same data is also used to signal suppliers to send the parts needed to build the product. With EDI, it becomes possible to reach the goal of "sell one, build one."

Besides making information available almost instantaneously in the supply chain, EDI also assures virtually 100% data accuracy. Accurate forecasts are vital to suppliers and the 830 transaction set (described on page 29) allows manufacturers to transmit forecasts directly into the production scheduling systems of suppliers. This enables suppliers to direct resources much more precisely to where they are needed and when they are needed.

What is the level of JIT awareness at your company? at your suppliers? at your customers?

_____

_____

_____

_____

_____

_____

_____

_____

Of the following EDI core competencies, in which ones do you need more education and training?

	More Education and Training?	
	Yes	No
JIT Principles		
Set-up Reduction		
Supplier Certification		
Statistical Process Control		
Total Quality Control		
Preventive Maintenance		
Value Analysis		
Problem-solving		
Team Building		

EDI Transaction Sets

The following are typical of the EDI transaction sets used in manufacturing.

Request for Quotation (RFQ)
ANSI X12 #840

Contains product specifications, quantities and delivery schedules.

Purchase Order (PO)
ANSI X12 #850

Contains all the information in a traditional purchase order. This is probably the most widely used transaction set at the present time.

Planning Schedule/Material Release
ANSI X12 #830

Transaction set for planning schedule contains anticipated needs for a set period of time in the future. Supplier uses information to plan own procurement and scheduling. Material release transactions are usually generated automatically by computer application. Alerts supplier of need to replenish. Specifies what is needed, how much and when and where to deliver.

Shipping Schedule
ANSI X12 #862

Depicts daily or in-sequence needs of a manufacturer. Alerts supplier to when and where to deliver according to JIT schedule.

Advanced Shipping Notice
ANSI X12 #856

Supplier's response to manufacturer's shipping schedule. Alerts manufacturer that parts are ready to be shipped.

Receiving Advice
ANSI X12 #861

Notifies supplier that shipment was received by manufacturer.

Payment Remittance Advice
ANSI X12 #820

Accompanies payment and contains information about payment history of original purchase order. Used for reconciliation.

Integration and EDI Strategy

Unlike one-shot methods at solving manufacturing problems, EDI is a tool which can help you take a strategic view of your company. EDI makes it possible for you to electronically integrate your company's supply chain — customers to manufacturers to suppliers. But you have to think of EDI as more than a way to transmit invoices and POs. This new and growing technology is a means of communicating and exchanging information so that the entire community can begin the process of reengineering current processes. And because EDI is fast and accurate, it is the best method for achieving this type of integration.

You need to integrate an EDI strategy into your company in order to maintain your competitive position in the global marketplace. The integration of customer orders with forecasts and delivery and production schedules requires that you have world class suppliers in this contest for the customer's dollar. In *Supplier Certification II: A Handbook for Achieving Excellence through Continuous Improvement* (0-945456-08-5; PT Publications; West Palm Beach, FL), Peter Grieco explains a number of ways of finding these suppliers.

Adopting a strategy of EDI integration means that bottlenecks and disconnects can be eliminated. You will also find that other manufacturing strategies such as set-up reduction, lot size reduction and cycle time reduction are also facilitated. Peter Grieco and Carl Cooper address cycle time reduction in their latest book, *Power Purchasing: Supply Management in the 21st Century* (0-945456-13-1; PT Publications; West Palm Beach, FL). The key is to coordinate internal processes with external ones as the following list demonstrates:

- **Process customer orders.**

- **Determine parts, materials and subassemblies required to fill the order.**

- **Determine production schedules to meet customer demands.**

- **Generate material releases (internal and external) to support production.**

- **Keep records of shipments.**

- **Make and document payments.**

- **Document use of parts, materials and subassemblies for inventory levels.**

EDI, as you can see, will help you achieve the goal of 100% customer satisfaction while making a profit. The extended community of the supply chain must use EDI to embrace philosophies and technologies such as JIT, bar coding, benchmarking, reengineering and Total Quality Management.

A Manufacturing Case Study

GE Aircraft Engines (GEAE) is at the center of a network that includes over 500 suppliers. Early on in their implementation of EDI, the aircraft engine manufacturer opted for a philosophy in which mere connectivity was not the goal. GEAE was looking at activities such as electronic PO transmissions as a step toward full integration, what the company calls Big EDI.

Since over two-thirds of the cost of their product can be traced to their supplier base, GEAE recognized that working with their suppliers was essential to realizing the full advantages of EDI. The company was looking to connect suppliers directly to the company's information flow so that there would be no nonvalue-added human intervention. This was more easily accomplished with larger suppliers who already had the hardware and personnel to take on this task. Smaller suppliers were not so fortunate until GEAE saw to it that they were included as well.

The manufacturer called on GE Information Services (GEIS) to assist it in helping these suppliers move away from being "rip-and-read" EDI users. A GEAE/GEIS team visited key suppliers and developed process maps of their information flow with the aim of finding an integrated solution. Once the solution was found, GEAE then began educating and training its suppliers in its use. The benefits are truly win/win. GEAE reduces cycle time and costs, and suppliers are able to survive and thrive in a competitive marketplace.

EDI
In
Retailing

Chapter Five

If any industry depends on its relationships with suppliers, it's the retail industry. EDI is an important ingredient in cementing these relationships and turning them into win/win situations. Retailers are looking for suppliers who have a clear vision of how to implement and integrate EDI technologies well into the future. Again, we see that the supplier with a business strategy has the best chance of competing and surviving. This strategy must integrate EDI so that retailers and suppliers can share data and applications.

In the past, suppliers have adopted EDI because large retailers like grocery chains and mass merchandisers have demanded integration as a condition for doing business. The time has come, however, for suppliers to take a more proactive approach. Such an approach will often entail reengineering efforts aimed at streamlining business processes to take advantage of EDI. Some of the facets of EDI which suppliers and retailers use today are:

- **E-mail.**
- **Electronic Funds Transfer.**
- **Online Catalogs.**
- **Internet.**
- **Bulletin Board Systems.**

Each of these facets works in synergy with EDI and helps to effect the integrated applications which make EDI a win/win proposition for suppliers and retailers. The synergistic elements of this relationship are:

- **Customer Satisfaction.**
- **Quick Response Systems.**
- **Immediate Replenishment.**

Now we are talking the language of proactivity. Let's turn our attention to an important link in the supply chain — logistics — and how EDI can be used to improve the process.

EDI and Logistics

At first glance, logistics seems to be all about getting products from one place to another, but there is another layer that is not readily apparent. Logistics is also about moving massive amounts of data from one location to another. Retailers can use this data to bring carriers into the EDI loop that services the supply chain. They can do so with an eye toward improving customer service. An improved competitive position in retailing is dependent on the fast and accurate shipment, distribution and storage of parts.

A number of EDI transaction sets are important to the logistics area of retailing — Advance Shipping Notice #856, Motor Carrier Shipment Information #204, Motor Carrier Shipment Status Message #214 and Motor Carrier Freight Details and Invoice #210. Suppliers who use these EDI tools can now send loads and pay for deliveries much more quickly and accurately. That is why it makes economic and strategic sense for retailers to find suppliers who are EDI-capable. They should be able to contract business electronically using the four transaction sets listed above which are described in more detail here:

Advance Shipping Notice
ANSI X12 #856

Allows supplier to scan product labels and send shipment information before delivery. Retailer can then receive shipment more quickly and accurately.

Motor Carrier Shipment and Invoice
ANSI X12 #204

Functions as electronic bill of lading. Often contains more useful information than found on traditional paper documents.

Motor Carrier Shipment Status Message
ANSI X12 #214

Used by carrier to transmit information to supplier and retailer. Notifies both of changes in shipment status or alerts them to reroutings, etc.

Motor Carrier Freight Details and
Invoice
ANSI X12 #210

Contains information on shipment changes and invoice for shipment. Used to reconcile freight invoice and bill of lading. Permits invoice to be paid automatically.

What steps can you take to improve your logistics function with EDI?

_____

_____

_____

EDI at Grocery Retailers

Let's now turn our attention to how EDI works in a specific business environment, that of a grocery retailer. The procurement cycle, as Figure 5-1 demonstrates, begins with the generation of a purchase order by the retailer. The information in the purchase order is converted to the Uniform Communication Standard (UCS) and then sent either directly to the grocery supplier or to a third-party Value

Added Network (VAN) which forwards the electronic PO to the supplier. In either case, the supplier receives the PO and verifies against internal customer files for credit. At the same time, the supplier's computer applications check inventory files for product availability. The order is then sent to the supplier's warehouse or distribution center where

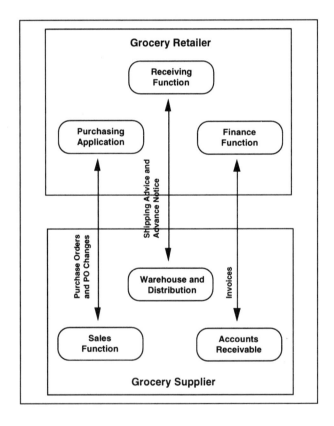

Figure 5-1.

the order is picked and readied for shipment. Advance notice of this shipment is then sent to the retailer via EDI.

Shipping information is forwarded to the finance function of the supplier to generate an invoice which is electronically transmitted to the retailer. While this is happening, EDI is also used to update or adjust the PO. It should be remembered as well that, ideally, each of these transactions would occur automatically as a result of both the supplier's and retailer's computer software.

EDI Implementation at Mass Merchandise Retailers

As you may expect, EDI has been of critical necessity to mass merchandisers who must cut internal costs to stay profitable. With their huge quantities of products and proliferation of suppliers, all retailers find that an EDI program is a massive undertaking. Let's look at some guidelines for how they conduct the process of implementation. The principles they have developed are applicable to a wide range of companies.

Principles of EDI Implementation

- **Set goals and define objectives. Most retailers want to reduce lead time, costs and errors.**

- Select and court suppliers capable of readily adopting EDI. They should represent a significant proportion of your volume of business.

_____

_____

- Determine what resources you can budget to the EDI process. It's better to know your limits before you start than to encounter them halfway through.

_____

_____

- Evaluate the way you currently conduct paper transactions and get rid of nonvalue-added activities.

_____

_____

- Establish clear procedures and documentation for your suppliers. This should cover both implementation and performance measurements.

_____

_____

- Work with your suppliers to establish time frames and mutually beneficial (win/win) partnerships.

_____

_____

EDI and Apparel Retailers

Apparel retailers have been leaders in the integration of bar coding, point-of-sale systems and EDI. As with other industries, large manufacturers often are the drivers of EDI adoption. For example, the jeans manufacturer Wrangler recently implemented a plan for approximately 1,000 of its independent retailers to restock their shelves by scanning a bar code. Under this system, stores send daily sales reports directly to Wrangler which the manufacturer combines with a store profile and seasonal adjustments to make a shipment. Retailers are given flexibility in this plan to add or subtract items from the shipment as they deem necessary.

This system, or similar ones, are in use throughout the apparel retailing industry. They address the most critical element of retailing — inventory management — with an automatic or semiautomatic system which uses electronically transmitted data. The obvious advantage, once again, is the combination of speed and accuracy which leads to reduced costs and cycle times.

EDI Associations for the Retail Industry

We recommend contacting two national associations

for those companies interested in obtaining industry guidelines and standards. The first is VICS (Voluntary Interindustry Communications Standards) which sponsors a conference exploring the integration of EDI and bar coding. The second organization is UCC (Uniform Code Council) which works with VICS and sponsors user group conferences.

EDI
AND
PURCHASING

CHAPTER SIX

EDI is still in the process of shaping the purchasing department. Here are some trends that we think are important to understand. The potential of this technology to transform the way we purchase products and services is great and cannot be ignored. EDI is one of the most important components of the wave of the future.

Open-EDI

The goal of the purchasing department has always been to promote efficiency in the purchase of products and services. Today, we use a combination of paper, multipart POs, faxes and telephones to make purchases from suppliers. If EDI is to become the future way we conduct business, then we need the ability to conduct business spontaneously and securely. This type of electronic purchasing is called "open-EDI."

To be truly open, EDI must include international links as well. Simply having spontaneous and secure links between two businesses or a group of businesses in one industry is not enough. A disk manufacturer in Malaysia needs to be able to link electronically with a chip maker in California.

The problem with open-EDI is not one of translation and connectivity. The technology is there, but small companies don't have the leverage or technical expertise to put it into effect. And large companies because of their bureaucracy and culture move very slowly in putting these technologies to work.

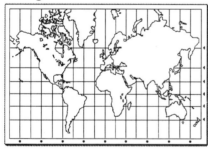

**Open-EDI—Spontaneous
and Secure Links
Around the World**

How would open-EDI help your company?

_____

_____

_____

_____

Let Your PC Do the Purchasing

Thomas Publishing, publishers of the Thomas Register, have begun a service known as ConnectsUs which allows companies to make purchases electronically. ConnectsUs is an electronic catalog on the General Electric Information Services' network. Purchasing personnel can use this catalog to make selections and place an electronic order. ConnectsUs converts the purchase order to EDI formats and GE's network delivers it to the computer at the supplier's place of business. The electronic purchases themselves are made with a purchasing card number.

With systems like this, it is pos-

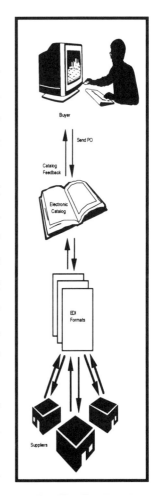

sible to look up all the suppliers of a particular product or service and then to compare features and prices. Suppliers are also able to update or add to their electronic catalogs much more quickly and at a far lower cost than with print methods or even CD-ROMs.

EDI, Purchasing and the Web

One of the growing areas of electronic commerce is the World Wide Web, or simply the Web. EDI has made its presence felt here as well. One company, iWTC, has developed a Web site which functions like a skyscraper. Electronics companies, for example, are on 11th floor and government offices on the 15th. A user simply clicks on the appropriate icon with a mouse and finds himself in the Web site of the selected company. There, he can search through catalogs provided by the electronic skyscraper's occupants. iWTC supports order entry and EDI for those wishing to make a purchase.

Do you have an Internet connection? If you don't know, who should you ask in your company to find out?

_____

_____

How would you rate your knowledge of the Internet and the World Wide Web on a scale of 100?

_____

If you rate yourself at less than 85, where will you go to find more information and help?

_____

_____

EDI and Rail Transportation

In another use of EDI, two rail transportation companies have improved communications and become more efficient as a result. This has had repercussions even for businesses not directly involved since more efficient operations lead to lower costs.

Conrail and Canadian National now exchange billing and rate information via EDI about freight shipments between their lines. The EDI system replaced a tedious and expensive system using paper. The new system improves accuracy and the timeliness of bills sent to shippers. There is no more manual communication, receiving, interpreting, or rekeying, which had contributed to a number of errors.

List areas in your company where EDI could improve communication about transporting goods.

_____

_____

_____

_____

How much could your company save in terms of time, people, resources and money?

_____

_____

_____

_____

List areas in your company where EDI could improve the relaying of information about the delivery of services.

_____

_____

_____

_____

Supply Chain Management

EDI can be used to support activities throughout the supply chain in order to lower cost, improve response times, attain world class quality levels and achieve flexibility in product/service delivery.

For example, information from electronic POs can be used by packers. The system sends an ASN (advanced shipping notice) which they then use as they scan items.

Thus, the contents are automatically checked against the order.

All of this helps companies move toward smaller and more frequent shipments which can be delivered directly to the point they're needed. This is the essence of JIT and flexible manufacturing.

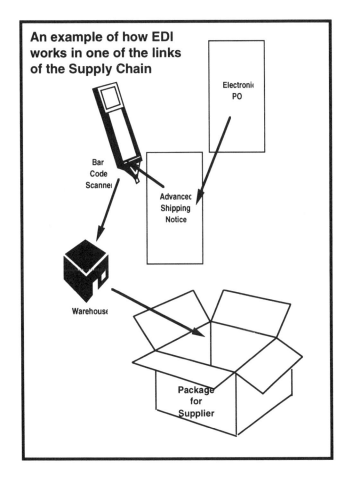

An example of how EDI works in one of the links of the Supply Chain

JIT Purchasing

With EDI and JIT, you can eliminate most of the paperwork involved in the purchasing process.

A combination of EDI and Kanban has been used by Motorola to determine what material was used by each assembly line and to signal the warehouse for replacement. Full EDI will be able to send this signal directly to supplier for each part used. Suppliers like this system because they know exactly how much to build to meet demand. Such a system can only be done with suppliers who have met supplier certification standards.

Benefits of Combining EDI and JIT

- Reduce supplier order cycle time

- Reduce warehouse space

- Increase inventory turns

- Receive on-time deliveries

- Reduce costs, while improving quality

• Permit automatic release of POs and delivery of goods directly to "point-of-use"

• Set the stage for closed loop EDI purchasing — from production plans to payment to supplier. No more requisitions, mailing of POs, or order entry

EDI Purchasing in the Military

Even the U.S. Army has an electronic purchasing program known as DVD/EDI (Direct Vendor Delivery/Electronic Data Interchange). Greatly simplified, the system works like this. A soldier needing to replenish an item

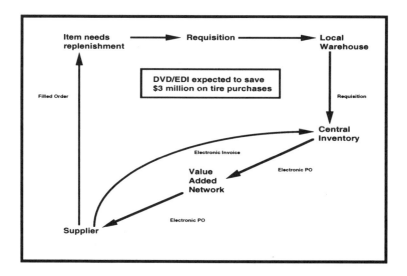

makes a requisition. If the local warehouse is out, the request is forwarded to a central inventory control point. This center then sends an electronic PO for the item to a VAN which forwards the order to a supplier. Once the supplier receives the PO and fills the order, the item is shipped to the Army unit. The supplier also sends an electronic invoice to the Army's inventory point which updates its files.

Using this EDI program, the Army expects its order cycle to reach two days. The Army expects to save $3 million alone in the purchase of tires for HMMWVs (High Mobility Multipurpose Wheeled Vehicle).

EDI and Purchasing Cards

Purchasing is always looking to control expenses and improve reporting. EDI is an important tool in this regard, although it has primarily been used for direct purchases. We believe that there is no reason that EDI can't be used for indirect purchases as well — office supplies, temp. services, etc. In fact, we estimate that indirect purchases account for 80% of a company's transactions and 20% of actual money spent.

One solution for bringing indirect purchases into EDI is the use of purchasing bank cards. Purchasing cards eliminate the need for requisition forms and petty cash. They allow the consolidation of invoices. Because authorization is built in, users can't charge above set limits. In addition, suppliers get prompt payment and customers get on-time delivery. All of these benefits directly complement EDI.

List indirect purchases for which purchasing cards could be used.

_____

_____

_____

_____

_____

_____

What type of authorization and limit would you need to set for a purchasing card at your company?

_____

_____

_____

_____

_____

_____

Real Time Purchasing with EDI

Purchasing professionals need to find products and services quickly and to order them electronically and have the ability to check price and quantity and obtain instantaneous updates. Buyers would also like to be able to do all of this while on-line. It's becoming more apparent every day that real time EDI (being able to do what is described above) is on-line EDI.

Real-time EDI thus requires a client/server architecture like that found on the Internet. This enables users to make a direct connection with a supplier and receive updates in real-time. This ability corrects what many see as a deficiency of VANs which store information and makes it accessible to users. With real-time EDI, purchasing professionals can make more informed decisions since they have the latest product information.

Here's how real-time EDI works at Wareforce, a computer products reseller. The customer makes up an order while off-line. She then connects to an on-line system where availability and prices are updated. If the item is out of stock, Wareforce's system suggests substitutes within pre-arranged limits. Then the order is placed with the customer knowing exactly what she is getting.

A typical transaction at a 14.4 baud rate takes 30 seconds, compared to phone calls which can take 3 to 5 minutes. The savings may not appear great, but when repeated hundreds of times, it soon adds up.

McJunkin, the pipe, valve fittings and electrical distributor, has a similar client/server

Typical Phone Transaction
3-5 minutes
Typical EDI Transaction
30 seconds

system which they believe is much more timely than a VAN's store-and-forward processing system.

EDI and BPR (Business Process Reengineering)

Companies have found that EDI is much more than paperless purchasing. It can be a catalyst for sweeping changes in the way business and internal processes are conducted. In short, it can be used as a BPR tool. EDI can help a company rethink its production and logistics, explore ways to bring products to market, and compete in the global marketplace.

How? By using EDI to design a system in which the business goals of both the customer and buyer are met. Some of these goals are to eliminate nonvalue-added activities, to ensure highest quality and lowest total cost, and to

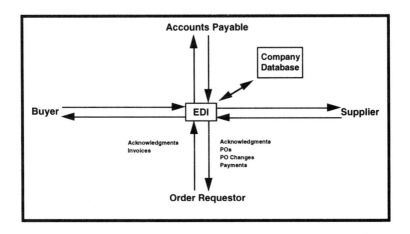

do it right the first time. EDI helps a company make a common database. Using this shared information, a company is now able to make connections between the buyer, accounts payable, and supplier. In addition, EDI supports feedback between each of these parts which will enable the company to seek even more improvement.

IMPLEMENTATION

CHAPTER SEVEN

We recommend that you use the same method for implementing EDI as you would for any other major business strategy. The main idea is not to do the implementation all at once, but as a planned sequence of steps. Before you begin, we also think it is wise to address two initial considerations — resistance and management buy-in. They are the keys to success, and they are often overlooked in the rush to start the implementation process. But it must be remembered that these two issues are also part of that process. More on this subject can be found in a very informative book that we have published: *People Empowerment: Achieving Success from Involvement* (0-945456-07-7; PT Publications; West Palm Beach, FL) by Wayne Douchkoff.

Resistance in Your Company

As your company begins to implement EDI Purchasing, there will be some resistance to this push for change in your organization. Overcoming this resistance is a challenge that is best met by educating people about the opportunities that EDI Purchasing represents. One of the benefits which should be stressed is the pride and sense of accomplishment which comes from helping to build and then work in a world class company. Beyond these morale boosters, however, your company should be prepared not only to have everyone participate in the challenges, but to share in the benefits.

When we have been called in to assess resistance at a company, we have discovered that most problems usually are the result of one of these conditions:

- Teams believe that management is out looking for a scapegoat, somebody or some group to blame for everything that is going wrong.

- People believe that the change process is just a way for management to check up on their performance so that people can be laid off.

- People know what the problems are and who is causing them, but they can't say because of loyalty to a fellow worker or fear of a supervisor or management.

- People have worked in a crisis management environment where change has come to represent out of control conditions. The resultant negative stress leads to behavior in which people avoid the source of change.

- "Teams are a waste of time. Management never listens to them. Besides, we tried this ten years ago and nothing changed ..."

Our strong advice is not to allow these attitudes and feelings to remain suppressed. Bring them out in the open and discuss them fully and honestly. Only in this way can you start to build the sense of trust between people and management which is so essential to the people involvement/empowerment process.

One method we have used with success is to list the above conditions and ask team members whether any or all are in effect at their company. Open discussion of these issues in a team setting is the most effective method of overcoming resistance to change. It is rare to have change successfully imposed from the upper levels of an organization. Change is easiest and most successful as a grass-roots process.

Understanding how people handle change is also important. Studies on how people deal with the death of a loved one show that there are five stages that one must go through in order to deal with the death. Psychologists theorize that people go through similar, if not the same, stages when dealing with change in the workplace. The acronym "SARAH" is a memory aid for these five stages.

S = SHOCK
A = ANGER
R = REJECTION
A = ACKNOWLEDGMENT
H = HOPE or HELP

Management needs to recognize these phases and encourage people to transition through them in order to become empowered.

One of your primary responsibilities is to integrate EDI into the company without upsetting people. Again, people are most afraid of what they don't know about, so be open and up-front above all else. With the introduction of EDI, robots, automation and bar coding, the factory of the future will be highly technologically oriented. Job security becomes the number one concern of people and the problem in most need of being addressed. People need to be assured that their jobs will not be eliminated. There will certainly be different kinds of jobs and the company should provide education and training for any transition.

All of the above ideas must be implemented to smooth the transition, but, once again, attitude is the most crucial factor. If the changes and even the techniques to smooth the way for the changes are put into place without the involvement of people who are empowered to make their own choices, then the integration of any new technology or idea will be strongly resisted. People begin to lose touch (sometimes literally because of automation) with their jobs and the results can be lessened motivation, powerlessness and resentment.

As we have said all along, the time has come to get rid of the attitude which says that workers have no brains and that management knows what is best for them. Despite what Frederick Taylor may have learned from his time-motion studies about speeding up work, more recent psychological studies show that people who are deprived of a say in their work display the same characteristics mentioned above — lessened motivation, powerlessness and resentment.

Write down five ways in which you will overcome resistance.

_____

_____

_____

_____

Resistance from Trading Partners

Resistance will not only come from within your company or purchasing department. Your trading partners' reactions will range from acceptance to skepticism. There are two reasons why trading partners resist EDI implementation:

- Lack of motivation.
- Lack of ability.

Suppliers see that EDI changes the way business will be conducted and that makes them afraid. Your job is to manage this change by influencing how EDI Purchasing is perceived and understood. Here are some suggestions that we have found successful:

- Schedule supplier conferences.
- Show a video on EDI purchasing.
- Confront the issues head on.
- Demonstrate real-world uses.

Nothing tops giving people hands-on experience. We advise actually sitting them down at a terminal and letting them order something. When they see that it can be done in a couple of minutes instead of a couple of days, they will be convinced. And motivated.

> **Write down five ways in which you will overcome resistance from your trading partners.**
>
> _____
>
> _____
>
> _____
>
> _____

Management Buy-In

No implementation of EDI Purchasing is complete without management commitment. They must provide the money, resources and people to make the process a successful one. We recommend that you approach management in much the same way as you would a trading partner. Prepare a cost/benefit analysis to show how EDI Purchasing will save money, improve quality and increase flexibility.

Once management has made a commitment, their job is to set the following activities into motion:

- Establish a strategic vision.

  Who? ─────────────
  When? ─────────────

- Set goals and objectives with specific start and end dates.

  Who? ─────────────
  When? ─────────────

- Set goals and objectives which can be easily measured.

  Who? ─────────────
  When? ─────────────

- Find leaders, champions, sponsors to guide the process.

  Who? ─────────────
  When? ─────────────

- Empower teams and employees.

  Who? ─────────────
  When? ─────────────

- Get outside support from BPR and EDI experts.

  Who? ─────────────
  When? ─────────────

Establishing a Vision

A strategic vision for what EDI purchasing is expected to accomplish is of utmost importance. But what goes into the creation of a company's vision statement? What are the common ingredients which allows companies like yours to implement successful programs? From our experience in the field, we have been able to isolate five factors:

- **The commitment of top management.** This is where the formulation of a vision must start. That vision must be driven by the marketplace in concert with the strategic plan. The vision must not be one handed down from on high. It should be more like a set of rules whereby people can take responsibility and be granted the authority to create a better company.

- **Total participation of management.** Never expect people to engage in a process of which you aren't willing to take part. Management must realize that they are a service organization for the employees.

- **Participation by people.** The people who work at a company are every bit as important as management. They should be listened to.

- **Continuous improvement teams.** This is the grass-roots action level. Without teams, the process would become a management-led program. As we have found out in the past, such programs are never as effective as ones that are activated by people.

One effective way to help people become aware of a company's vision is to allow them to participate in the development of statements which convey the mission of both the company and its people. The involvement in the development of the vision acts as a motivator in making the success of the EDI Purchasing a primary responsibility of all people in your company.

Who are the champions of EDI at the management level in your company?

_____

_____

_____

Who are the members of management who will need to be educated?

_____

_____

_____

When are you going to begin your education campaign and who will be on the team? Who will lead the team?

_____

_____

_____

Strategic Plan

We have been talking quite a bit about strategy, so let's focus on what we mean. What do you need to do to fashion a strategic plan for the implementation of EDI purchasing? First, you must consider where the overall business of your company is headed. What are the business needs of your organization? What is your company's tolerance for risk?

Then, you need to look at how your competition uses EDI in the purchasing department. Also, take a look at the use of EDI in your industry or service group. With this information, you can then begin to evaluate the financial limits of your company with respect to its future needs.

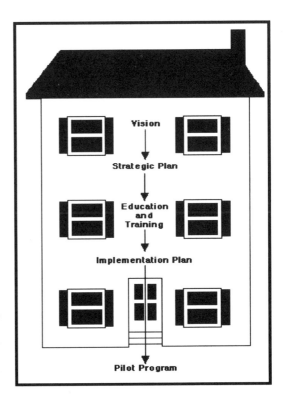

Most importantly, don't forget to ask yourself what your

customers want.

Lastly, you should assess your current EDI capability and determine how much education and training is needed and at what levels.

The strategic plan itself requires the goal setting and vision creation mentioned above. It also involves developing the infrastructure of your company. What new computer and telecommunication systems will you need in order to implement EDI purchasing? Another essential part of the plan is obtaining early supplier involvement. Trading partners must be part of the team effort.

Now that the foundation has been built, the EDI Purchasing team is ready to create an implementation plan for a pilot program. The team is now able to work out the details with the education and training you have provided them.

Implementation Plan

The implementation plan is the product of the EDI Purchasing team. The first job of this team is to gather information about both the successes and difficulties of implementing EDI in the purchasing department. This may entail joining industry associations which deal with EDI. Another source of information can be had by evaluating EDI providers of hardware and software, as well as consultants who can provide services. And, of course, find EDI users and have them show you how the system works in their purchasing department.

The next step for the team is to analyze the information and process flows in the purchasing department. At the same time, the team should be determining which suppliers (trading partners) are most conducive to change. This is also the time to determine the costs of implementing an EDI system.

We then recommend that you plan a pilot program with one of your trading partners. Remember to set up a system to assess your progress as this pilot begins. The information you gain can be used to help you clone the program with other trading partners.

Step	Goal	Due Date
Information gathering	_____ _____	_____
Information and process flow analysis	_____ _____	_____
Supplier selection	_____ _____	_____
Pilot program	_____ _____	_____
Measurement and assessment	_____ _____	_____

One company which has been successful with their EDI implementation is Home Depot. They were able to do so because they paid close attention to two main considerations:

- Developing a trading partner agreement which is specific as to who does what and making it a win/win document.

- Starting with suppliers who are EDI capable so you can learn from their experience.

These are important considerations which will prove to be valuable in your implementation as well.

Notes:

_____

_____

_____

_____

_____

_____

_____

_____

_____

_____

_____

_____

_____

_____

Staples Online EDI Implementation Guide

In another example of the usefulness of the Web, Staples, the office supplier headquartered in Framingham, MA, has posted an online implementation guide showing suppliers how to do business with the company using EDI.

The Table of Contents on Staple's Web page shows what is covered:

STAPLES EDI IMPLEMENTATION GUIDE

REVISION 003
02/10/95

TABLE OF CONTENTS
* Important Names and Numbers
* Staples Location Information
* Staples EDI Transaction Sets
* Electronic Commerce at Staples
* Staples Terms/Conditions for Purchase Order
* Staples Virtual Office Superstore

Following the Table of Contents is an introductory document entitled "Getting Started with Staples." The document covers quite a bit of material, but here are some of its highlights:

• Identifies the VAN they use — Sterling Software's COMMERCE:Network.

- Instructs potential users about what to do if they are already a member of this VAN and what to do if they aren't.

- Notes testing which is done to get system working when you become a trading partner.

- Tells what document types and versions Staples supports.

- Shows other networks they are connected to such as GE, Sprint/Telnet, Advantis, etc.

- Talks about how to use a VAN.

- Notes translator they use and what translators are compatible with their system.

- Offers to help and educate companies wishing to implement EDI.

As you can see, the guide addresses hardware, software and systems questions, as well as providing potential users with the steps needed to take advantage of EDI Purchasing with Staples.

SELECTING SOFTWARE

CHAPTER EIGHT

This chapter consists of three audits. The first helps you determine how ready you are to implement EDI. The second will help you determine the best software supplier for your EDI needs. And the third audit is a list of guidelines for evaluating the software itself. It almost goes without saying that computer software is the core of EDI, but too many companies ignore the rigorous process necessary to find the software best suited to their needs. EDI implementation is not as easy as going to your local computer store and taking a box off the shelf. Furthermore, we recommend that these audits be conducted by a team consisting of a broad-based group of users who will be affected by EDI.

EDI Readiness Audit

Answer questions on a scale of 1 to 10
with 10 being the highest score.

- How would you rate your general understanding of how EDI works?

  Points ____

- Are you able to define the cost benefits and reductions that your company will accrue from EDI?

  Points ____

- Have you and a team written a comprehensive report outlining your company's plan for EDI to be presented to senior management?

  Points ____

- Rate your understanding of the technical issues involved in EDI? Can you make yourself understood to both everyday users and to experts?

  Points ____

- What type of leader are you? Can you educate people to overcome obstacles to EDI implementation?

  Points ____

- What are the chances that senior management will allot the resources you will need to bring EDI online?

  Points ____

- How complete is your implementation plan? Does it show the impact on your company?

  Points ____

- What are the group skills of your EDI team? How well do they know brainstorming, scheduling, value analysis, team building and cost justification techniques?

  Points ____

- How complete is your performance measurement system? Will it keep you on track and alert you to problems?

  Points ____

- Rate your understanding of industry standards and specifications. Do you have written documentation covering this area?

Points ____

Total Points ____

Score   **Comments**

95-100   You are ready to begin implementation.

80-95   Take time to gather more information or get help from third-party consultants.

below 80   Take time to find out what is missing. Look for seminars and conferences where you can obtain help.

Selecting an EDI Software Supplier

Score 2 points for each "Yes" answer.

Category 1: Quality and Customer Satisfaction

YES   NO

1. Does this program have procedures for identifying, tracking, managing and reporting changes and revisions?   __   __

2. Does the supplier have a backup and security system in place for procedures, instructions, computer systems, files and documents?   __   __

3. Does supplier have a security system for protecting customer software, present software work, files and documents?   __   __

4. Does the supplier have long-term experience in partnership agreements with customers? __   __

5. Does management regularly follow up on the progress of the software quality improvement plan?   __   __

6. Has the supplier's top management released a written software quality policy, quality values or quality commitment?   __   __

7. Does the supplier have a documented software quality assurance system?   __   __

                                                    YES   NO

8. Does the supplier have procedures, instructions and mechanisms for applying quality plans to software work?   —   —

9. Is there a follow-up system for software quality metrics and corrective actions?   —   —

10. Has the supplier defined a procedure for evaluating critical factors in software for reliability, safety, standards and performance?   —   —

11. Does the supplier have a system for maintaining customer software after delivery?   —   —

12. Are recovery plans in place in case of disasters?   —   —

13. Are "open communication standards" supported by the software?   —   —

14. Does the software have a formal means of adding trading partners?   —   —

CATEGORY 2: Project Management

		YES	NO

15. Does the supplier have a software project planning procedure and is it followed? ___ ___

16. Does the supplier regularly use software project planning tools? ___ ___

17. Are software development projects planned formally and consistently? ___ ___

18. Does the supplier have a system for controlling information during the project? ___ ___

19. Are software project management metrics used to track project progress? ___ ___

20. Does the staff use project control mechanisms and tools? ___ ___

21. Does the supplier identify key risks to minimize their effects? ___ ___

22. Does the system ensure that the software is ready in time and fulfills all customer requirements? ___ ___

CATEGORY 3: Software Development Process

YES   NO

23. Does the supplier have policies and procedures for new software development? Do they result in clearly defined project plans with appropriate measurables and approvals?   —   —

24. Do software projects employ phasing and are the phases formally selected?   —   —

25. Is the customer able to include its requirements in the planning of new software?   —   —

26. Does the supplier have a system for translating customer requirements into design requirements?   —   —

27. Does the supplier document software requirements and update them regularly? —   —

28. Is a tracing system used for software projects from requirements to design and code?   —   —

YES   NO

29. Does the supplier have a system for ensuring coverage of the requirements?   __   __

30. Are all outputs of the phased development approach defined and fulfilled?   __   __

31. Does the supplier use approved design standards in software projects?   __   __

32. Do the standards cover software requirements, design, coding, documentation and testing?__   __

33. Does the project team routinely maintain software design reviews?   __   __

34. Does the approval procedure base entry and exit criteria on customer requirements?   __   __

35. Is there a documented process for developing software configuration?   __   __

36. Does the supplier have a procedure for managing software releases?   __   __

YES   NO

37. Does the supplier have a procedure for managing software changes?   —   —

38. Is there a system to notify project participants and customers of changes and to obtain approval?   —   —

39. Does the supplier employ software verification activities?   —   —

40. Does the supplier have a procedure for controlling sensitive hardware and software materials in all phases of development and during processing?   —   —

CATEGORY 4: System/Acceptance Testing

YES   NO

41. Does the procedure to control system/ acceptance testing guarantee sufficient levels of coverage and documentation?   —   —

42. Is software testing activity independent?   —   —

YES NO

43. Does the supplier perform testing under conditions similar to actual customer operating conditions?   —   —

44. Are the customer's operational profiles used to define test runs?   —   —

45. Does the supplier have a procedure for the formal acceptance of all phases of software development?   —   —

46. Does the supplier measure the effectiveness of its software test process?   —   —

CATEGORY 5: Documentation and Training

YES NO

47. Does the system provide audit trails?   —   —

48. Does the software provide you with flexibility in naming standards?   —   —

49. Does the supplier provide in-house training for technical, operations and user staff?   —   —

50. Does the supplier provide in-house
    training for management?   —   —

Score	Comments
95-100	You are ready to begin implementation.
80-95	Take time to gather more information or get help from third-party consultants.
below 80	Take time to find out what is missing. Look for seminars and conferences where you can obtain help.

Software Evaluation

A number of software companies are developing products for use in EDI. Before you buy software, we suggest that you find out if the program has the following features:

- Flexibility — has the capability of allowing the user to change the parameters in order to build a system suited to your particular requirements.

- User-Friendly — simple, intuitive menus and graphic interface which allows users to quickly learn the program.

- Industry Standards — program should be based on industry standards.

- Traceability — tracks reject actions and lists reason codes and dates for evaluations.

- Ease of Implementation — easily configurable with present equipment.

- Measurements — able to predict trends based on data, history and benchmarking; this should be a preventive method.

- Quality — ability to track trends for continuous improvement.

SELECTING A VAN

CHAPTER NINE

What are VANs? VANs are a powerful and practical solution to overcoming obstacles in direct communication via computer with trading partners. These obstacles to the efficient functioning of EDI Purchasing are listed here:

- Different communication protocols.

- Varying transmission speeds.

- Simultaneous phone line availability.

- Compatible hardware.

This is what VANs do to eliminate the above problems:

- Translate different protocols so that companies can communicate with each other.

- Convert varying line speeds so that EDI messages can be sent and received at line speed of sender or recipient.

- Maintain mailboxes for members so that they can access when they have time. No need for simultaneous connection.

- Hardware is immaterial. Everything is converted to standard interface inside VAN, then translated.

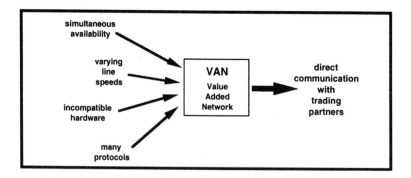

Advantages and Benefits of a VAN

The greatest advantage of a VAN is that it eliminates the need to reconfigure your system every time you initiate a new communication with a different trading partner. A VAN provides a single channel for you to call up. The VAN takes care of all the other channels connecting you to trading partners. To put it simply, a VAN reduces the number of connections you have to make.

In addition, the software, hardware and staff of a VAN handle all the technical areas of EDI translation. Another advantage is that there are now connections between all the major VANs which allows for almost world-wide EDI capability. Lastly, VANs are secure. Users need validation which means that nobody can make an order in your name or intercept a transmission directed to your company.

Do any of your trading partners use VANs? Which ones?

_____

_____

_____

_____

Which VANs are used in your industry or service group?

_____

_____

_____

_____

What technical areas will you need to address?

_____

_____

_____

_____

How to Select a VAN

The checklist below will help you make an intelligent choice. Select a VAN that has the following capabilities:

- Supports EDI transaction sets and standards used in your company's industry or service group.

- Connects to all other major VANs.

- Carries a large number of clients themselves which provides you with a large number of potential trading partners.

- Has knowledge about starting up and maintaining EDI Purchasing sites.

- Has a support staff which is available during all hours of operation.

- Has a knowledge of your business and its special needs.

- Practices what it preaches.

There is an excellent article in *EDI World*, February 1995, pages 32-39, on VAN selection. It contains a features grid with contacts and phone numbers of major providers.

Start Date for VAN Selection Team: _____

Team Leader: _____

Mission: _____

_____

_____

_____

_____

RESOURCES

CHAPTER TEN

User Groups

User groups are an effective way to discuss EDI Purchasing with people who have already attempted projects in their own companies. The groups give you an opportunity to learn from companies from a number of different industries and services. In addition, you can make very beneficial contacts and relationships in such groups. A good place to start in your search for a local user group is in the May 1994 issue of *EDI World* (pages 38-39). The magazine updates and publishes this list of regional user groups on a regular basis.

Do this task by: _____

Responsible person/team: _____

Comments: _____

_____

_____

Online Services

Look in the business areas of the major online services such as America Online and Compuserve. There you will find discussions of many facets of purchasing and technological innovations such as EDI. Since the discussions change often, you must visit the site yourself to see what is there.

The EDI provider, Premenos, has a very informative site on the Web. Their address is http://www.premenos.com. There you will find information about EDI standards and other electronic commerce resources.

Do this task by: _____

Responsible person/team: _____

Comments: _____

_____

_____

University Resources

Many universities and colleges have resources devoted to helping companies implement new technologies. Call business departments for more information. Central Connecticut State University, for example, has a Flexible Manufacturing Networks Center which has recently implemented an Electronic Commerce Center to provide EDI service to small manufacturers in Connecticut and beyond.

Do this task by: _____

Responsible person/team: _____

Comments: _____

_____

_____

HELP DESK HOTLINE: 1-800-272-4335

In order to answer the questions of our readers, we have established a Help Desk Hotline at our corporate headquarters in West Palm Beach, Florida. We invite you to call us with your queries about how to use the forms and tools in this book.

We also invite you to use our HELP DESK HOTLINE to find out more about *The Purchasing Encyclopedia* and other books we publish as well as a videotape series entitled *Supplier Certification: The Path to Excellence.* In addition to books, software and videotapes, we offer over 100 courses which can be scheduled for intensive, in-house seminars. Call us for details.

Do this task by:_____

Responsible person/team:_____

Comments:_____

_____

_____

Additional Purchasing Resources from PT Publications, Inc.

3109 45th Street, Suite 100

West Palm Beach, FL 33407-1915

1-800-272-4335

THE PURCHASING ENCYCLOPEDIA

Just-In-Time Purchasing: In Pursuit of Excellence
    Peter L. Grieco, Jr., Michael W. Gozzo
    & Jerry W. Claunch
	$29.95

Glossary of Key Purchasing Terms, Acronyms,
    and Formulas
    PT Publications
	$14.95

Supplier Certification II: A Handbook for
    Achieving Excellence through Continuous Improvement
    Peter L. Grieco, Jr.
	$49.95

World Class: Measuring Its Achievement Peter L. Grieco, Jr.	$39.95

Purchasing Performance Measurements: A Roadmap
    For Excellence
    Mel Pilachowski
	$12.95

The World Of Negotiations: Never Being a Loser Peter L. Grieco, Jr. and Paul G. Hine	$39.95
How To Conduct Supplier Surveys and Audits Janet L. Przirembel	$14.95

Supply Management Toolbox: How to Manage
    Your Suppliers
    Peter L. Grieco, Jr.
	$26.95

Purchasing Capital Equipment Thomas E. Petroski	$14.95

Power Purchasing: Supply Management in in the 21st Century Peter L. Grieco, Jr. and Carl R. Cooper	$39.95
Global Sourcing Lee Krotseng	$14.95
Purchasing Contract Law, UCC, and Patents Mark Grieco	$14.95
EDI Purchasing: The Electronic Gateway to the Future Steven Marks	$14.95
Leasing Smart Craig A. Melby and Jane Utzman	$14.95
MRO Purchasing Peter L. Grieco, Jr.	$14.95

CONTRACT MANAGEMENT SERIES

The Complete Guide to Contracts Management For Facilities Services John P. Mahoney and Linda S. Keckler	$18.95
The Complete Guide to Contracts Management For Components John P. Mahoney and Linda S. Keckler	$23.95
The Complete Guide to Contracts Management For Promotional Services William F. Badenhoff and John P. Mahoney	$18.95
The Complete Guide to Contracts Management For Business Practices William F. Badenhoff and John P. Mahoney	$23.95
The Complete Guide to Contracts Management For Office Services John P. Mahoney and William F. Badenhoff	$16.95
The Complete Guide to Contracts Management For Peripherals John P. Mahoney and William F. Badenhoff	$23.95

The Complete Guide to Contracts Management For $14.95
    Capital Equipment
    John P. Mahoney and William F. Badenhoff
The Complete Guide to Contracts Management $16.95
    For Human Resources Services
    John P. Mahoney and Linda S. Keckler
The Complete Guide to Contracts Management For $16.95
    Security Services
    William F. Badenhoff and John P. Mahoney
The Complete Guide to Contracts Management For $23.95
    Contract Manufacturing
    John P. Mahoney and William F. Badenhoff
The Complete Guide to Contracts Management For $18.95
    Distributors
    John P. Mahoney and William F. Badenhoff
The Complete Guide to Contracts Management For $18.95
    Transportation and Logistics Services Volume 1
    John P. Mahoney and Linda S. Keckler
The Complete Guide to Contracts Management For $18.95
    Transportation and Logistics Services Volume 2
    John P. Mahoney and Linda S. Keckler
The Complete Guide to Contracts Management For $16.95
    Travel Services
    John P. Mahoney and Linda S. Keckler

PURCHASING VIDEO EDUCATION SERIES

Supplier Certification The Path to Excellence
    Tape 1: Why Supplier Certification? $395.00
    Tape 2: Quality at the Supplier $395.00
    Tape 3: How to Select a Supplier $395.00
    Tape 4: Supplier Surveys and Audits $395.00
    Tape 5: Supplier Quality Agreements $395.00
    Tape 6: Supplier Ratings $395.00

Tape 7: Phases of Supplier Certification	$395.00
Tape 8: Implementing a Supplier Cert. Program	$395.00
Tape 9: Evaluating Your Supplier Cert. Program	$395.00
Complete Nine Tape Series	$1,995.00

PURCHASING AUDIO TAPES

The World of Negotiations: How to Win Every Time $39.95

PURCHASING SOFTWARE

Supplier Survey and Audit Software $395.00
Developed by Professionals For Technology , Inc.

ContractWare™
Developed by The Leadership Companies, Inc.

Business Practices	$599.00
Capital Equipment	$599.00
Components	$599.00
Peripherals	$599.00
Contract Manufacturing	$599.00
Distributors	$599.00
Facilities Management	$599.00
Human Resources	$599.00
Office Services	$599.00
Promotional Services	$599.00
Security Services	$599.00
Transportation and Logistics	$599.00
Travel Services	$599.00

Site License (unlimited users per site)   Call
Corporate License (unlimited users, unlimited sites)   Call
Administrative Library Database
   (requires site of corporate license) Call

CyberBase™
Client Server Software containing all 14 contract families
Developed by the Leadership Companies, Inc.
Individual Server Licenses   Call
Corporate License (unlimited servers, unlimited users) Call
Additional Installations   Call

ADDITIONAL PROFESSIONAL TEXTBOOKS

Failure Modes and Effects Analysis: Predicting $39.95
  and Preventing Problems Before They Occur
  Paul Palady
Made In America: The Total Business Concept $29.95
  Peter L. Grieco, Jr. and Michael W. Gozzo
Reengineering Through Cycle Time Management $39.95
  Wayne L. Douchkoff and Thomas E. Petroski
Behind Bars: Bar Coding Principles and Applications $39.95
  Peter L. Grieco, Jr., Michael W. Gozzo and C.J. Long
People Empowerment: Achieving Success from Involvement $39.95
  Michael W. Gozzo and Wayne L. Douchkoff
Activity Based Costing: The Key to World Class Performance $18.00
  Peter L. Grieco, Jr. and Mel Pilachowski

LIST OF ACRONYMS

ANSI	American National Standards Institute
ASN	Advanced Shipping Notice
BPR	Business Process Reengineering
DVD	Direct Vendor Delivery
EC	Electronic Commerce
EDI	Electronic Data Interchange
EFT	Electronic Funds Transfer
JIT	Just-In-Time
MAC	Message Authentification Code
PC	Personal Computer
PO	Purchase Order
RFQ	Request for Quotation
SIG	Special Interest Group
UCC	Uniform Code Council
UCS	Uniform Communication Standard
VAN	Value Added Network
VICS	Voluntary Inter-industry Communications Standards

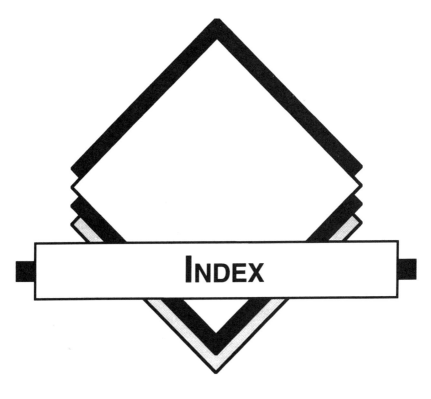

INDEX